I0427057

Table of contents

Chapter   Page Number

Introduction to Project Management for Engineers related to FGD

- The evolving role of project management in the engineering sector
- Overview of the book's objectives and structure
- Importance of effective project management in ensuring successful project outcomes

An Overview of Project Management in the FGD Sector for Engineers

The field of project management in engineering has experienced a notable transition, particularly in relation to Flue Gas Desulfurization (FGD) projects. Project management is essential to the success of FGD projects due to their dynamic and complex character, which necessitates a purposeful and well-coordinated approach. This introduction highlights the crucial necessity of competent project management in guaranteeing positive project outcomes, lays out the goals of this book, and sets the stage for understanding the evolving role of project management in the FGD engineering landscape.

Project Management's Changing Function in the Engineering Sector

1. **Complexity Adaptation:** Engineering project management has changed over time to meet the growing complexity of projects, especially in fields like FGD. Because FGD projects are complex and involve a variety of technologies, environmental factors, and regulatory compliance, a more advanced project management strategy is required.

2. **Exchange of Real-World Experiences**: One main goal is to exchange real-world experiences and lessons gained from the viewpoint of a Project Director working on FGD projects. Key ideas, difficulties, and effective tactics utilized during different project phases will be demonstrated through case studies and real-world examples.

3. **Advice for Project Managers and Engineers**: Engineers and project managers who are employed in the FGD industry or who aspire to work there will find the book to be an invaluable resource. It offers tools, best practices, and practical insights to improve project management abilities tailored to FGD projects.

4. **Integration of Interactive features**: The book includes interactive features including checklists, reflection activities, and talks on typical project management problems in the FGD industry to increase reader involvement. These components inspire readers to critically consider project management issues and actively apply ideas to their own projects.

5. **Integration of Diverse Stakeholders:** FGD projects in India typically involve multiple stakeholders, including government agencies, regulatory bodies, local communities, technology providers, contractors, and investors. Effective project management is essential for integrating these diverse stakeholders, aligning their interests, and facilitating collaboration towards common project goals.

6. **Navigating Complex Regulatory Environment:** India's regulatory environment for environmental projects like FGD can be complex and stringent. Project managers must possess a deep understanding of the regulatory requirements and ensure compliance throughout the project lifecycle. Failure to adhere to regulations can result in delays, fines, or even project shutdowns, highlighting the importance of effective project management in navigating regulatory hurdles.

7. **Managing Foreign Technology Partners**: FGD projects often involve the adoption of foreign technologies, with international technology partners playing a significant role. Effective project management is essential for managing relationships with these partners, ensuring effective

communication, addressing cultural differences, and mitigating risks associated with technology transfer and implementation.

8. **Ensuring Technical Integration**: FGD projects require the integration of various technical components, including desulfurization systems, scrubbers, monitoring equipment, and emission control systems. Effective project management ensures seamless coordination among engineering teams, technology providers, and contractors to ensure proper installation, testing, and commissioning of these components.

9. **Mitigating Risks and Uncertainties**: FGD projects are not without risks, including technical, financial, environmental, and regulatory risks. Effective project management involves identifying, assessing, and mitigating these risks proactively. It also involves developing contingency plans to address unexpected challenges, ensuring project resilience and continuity.

10. **Optimizing Resource Allocation:** FGD projects require significant financial, human, and material resources. Effective project management involves optimizing the allocation of these resources to ensure project efficiency and cost-effectiveness. This includes budgeting, procurement, scheduling, and workforce management to meet project milestones within budget and schedule constraints.

11. **Ensuring Quality and Performance:** Quality assurance and control are critical in FGD projects to ensure that systems meet performance standards and regulatory requirements. Effective project management involves implementing robust quality management systems, conducting regular inspections and audits, and addressing any non-conformities promptly to ensure that project deliverables meet or exceed expectations.

12. **Facilitating Communication and Transparency**: Effective communication is essential for the success of FGD projects, especially when multiple agencies and stakeholders are involved. Project managers serve as central points of contact, facilitating transparent communication among all stakeholders, providing regular updates on project progress, addressing concerns, and resolving conflicts to maintain project momentum and stakeholder confidence.

In conclusion, competent project management is essential to the success of FGD initiatives in India, which include several agencies and international technological partners. Through their ability to integrate a variety of stakeholders, navigate complex regulatory frameworks, manage technology partnerships, mitigate risks, optimize resources, ensure quality, and foster communication, project managers are essential to the success of projects and the delivery of environmental benefits that come with the use of FGD technologies.

The significance of proficient project management in guaranteeing triumphant project results

- **Reducing Risks in FGD initiatives**: FGD initiatives are inherently risky, with potential hazards ranging from changing regulations to technical difficulties. Proactive risk management throughout the project, early identification of any problems, and the creation of backup plans are all components of an efficient project management approach.

- **Maximizing Resource Utilization**: FGD initiatives frequently call for substantial human and financial resources. Optimizing resource usage, making sure budgets are followed, and assigning the appropriate people to the right tasks at the right time all depend on effective project management.

- **Fulfilling Regulatory Compliance**: Strict environmental rules apply to the FGD industry. A key function in navigating and guaranteeing adherence to these requirements is performed by project managers. Developing and implementing plans to meet and surpass regulatory criteria is made easier with the aid of a structured project management method.

- **Improving Stakeholder Communication**: Since many parties are involved in FGD projects, such as environmental authorities, local communities, and project investors, clear and effective communication is essential. Transparent communication is made easier by project management techniques, which also assist stakeholders develop trust and manage expectations.

- **Ensuring Project Delivery on Time**: Given regulatory deadlines and financial ramifications, time is frequently a crucial component of FGD initiatives. When projects are managed well, there are fewer delays and related expenses and projects are completed on schedule.

- **Promoting Continuous Improvement**: FGD initiatives provide insightful educational experiences. Continuous improvement is facilitated by project management techniques such as lessons learned sessions and post-project reviews. One project's lessons can be used to improve the efficacy and efficiency of subsequent FGD initiatives.

In conclusion, the evolution of project management in the field of FGD engineering is a challenging process marked by the integration of technology, global cooperation, sustainability as a priority, and flexibility in the face of complexity. This book highlights the critical role that effective project management, with its well-defined objectives and practical insights, plays in the success of FGD ventures. It provides engineers and project managers with the means to comprehend and implement project management principles, allowing them to seize opportunities and get past roadblocks in the dynamic domain of FGD projects.

Understanding FGD in respect to the Indian Engineering Landscape

- FGD and its importance related to Human Health
- Overview of the engineering industry in India
- Key sectors and projects

2015
- Original Notification issued by MoEF&CC (7th December 2015)
- Regulation to set limits for PM, So2 , NO2 & Hg emission and water consumption limits for the coal based power plants

2017
- First Extension in SO2 compliance timeline - CPCB issued direction under section 5 of EPA Act 1986, (11th December 2017)
- New timeline for plants in 300 Km radius of Delhi-NCR runs till December 2019

2018
- Dilution of Water Consumption Norms (28th June 2018)
- Water consumption limits for new power plants commissioned after 1st January 2017 were increased from 2.5 to 3 m3/MWh

2019
- Second Lapse of SO2 compliance by coal based power plants in 300 Km radius of Delhi - NCR (7th December 2019)

2020
- Jan-May 2020: Direction issued by CPCB for deposit of fines for illegal operation of power plants in 300 km of Delhi - NCR in absence of SO2 emission controls
- October 2020 : Emission norms for NOx for power plants installed between 2003-2016 diluted
- Second Extension of the timelines for Delhi -NCR plants by CPCB in Oct 2020 , staggered timeline running till December 2022

2021
- March 2021 : Third Extension for SO2 compliance granted to powr plants.
- Timeline extension till December 2024 , along with categorization of power plants units in category A,B & C

2022
- September 2022: Forth Extension for SO2 compliance granted to powr plants.
- Timeline extension till December 2027

Understanding FGD in respect to the Indian Engineering Landscape

<u>Story of Silent Suffering</u>

In the bustling city of Kolkata, amidst the cacophony of street vendors and honking vehicles, lived a man named Arjun. He was a hardworking father, striving to provide for his family in any way he could. Little did he know that the very air he breathed would become his silent assailant.

Arjun worked as a laborer near a coal-fired power plant on the outskirts of the city. Each day, he would brave the sweltering heat and thick clouds of smoke that emanated from the plant, unaware of the insidious danger lurking within. As he labored under the scorching sun, the toxic emissions of sulfur dioxide and nitrogen oxides enveloped him, seeping into his lungs with every breath.

At first, Arjun attributed his occasional coughs and bouts of fatigue to the grueling nature of his work. However, as time passed, his health began to deteriorate rapidly. His once robust frame grew frail, and his vitality waned with each passing day. Despite his worsening condition, Arjun remained stoic, unwilling to burden his family with worries about his health.

One humid evening, as Arjun returned home from work, he collapsed on the doorstep, gasping for breath. His family rushed him to the nearest hospital, but it was too late. The doctors informed them that Arjun had succumbed to a sudden respiratory failure, his lungs ravaged by years of exposure to toxic pollutants.

Devastated by their loss, Arjun's family mourned in silence, grappling with the cruel injustice of his untimely demise. They knew the truth behind his death, but they dared not speak out against the powerful entities responsible for it. In a country where industry held sway over justice, they feared the consequences of raising their voices.

And so, Arjun's death became just another statistic in the annals of India's industrial landscape—a tragic reminder of the silent suffering endured by countless individuals like him. As the coal-fired power plant continued to churn out its deadly emissions, unchecked and unchallenged, Arjun's family vowed to keep his memory alive, hoping that someday, their voices would break the silence and demand accountability for those who had perished in the shadows of industry.

Overview of FGD Impact

Amidst the bustling cities and industrial landscapes of India, the need to protect human health from the harmful effects of air pollution has never been more pressing. Flue Gas Desulfurization (FGD) emerges as a critical solution in this battle against toxic emissions, offering a ray of hope amidst the shadows of pollution and suffering.

- **Understanding the Threat**: The story of Arjun serves as a poignant reminder of the dire consequences of air pollution on human health. The toxic emissions from coal-fired power plants, laden with sulfur dioxide (SO_2) and nitrogen oxides (NO_x), pose a grave threat to respiratory health, leading to debilitating illnesses and premature deaths.
- **The Role of FGD:** Flue Gas Desulfurization (FGD) technology emerges as a beacon of hope in this bleak landscape. FGD systems, installed in power plants, act as guardians of the air, capturing and neutralizing harmful pollutants before they can escape into the atmosphere. By

removing sulfur dioxide and other noxious gases from flue gas emissions, FGD plays a crucial role in reducing air pollution and safeguarding human health.

- **Protecting Respiratory Health**: FGD technology directly addresses the primary sources of air pollution that contribute to respiratory illnesses like asthma, bronchitis, and respiratory failure. By significantly reducing the emission of sulfur dioxide and nitrogen oxides, FGD helps alleviate the burden of respiratory diseases, improving the quality of life for millions of people exposed to polluted air.

- **Preventing Premature Deaths:** The implementation of FGD systems in coal-fired power plants has the potential to save countless lives by preventing premature deaths attributed to air pollution. By curbing the emission of toxic pollutants, FGD technology mitigates the health risks associated with long-term exposure to polluted air, thereby extending lifespans and enhancing overall well-being.

- **Promoting Environmental Justice**: FGD technology not only protects human health but also advances environmental justice by ensuring that vulnerable communities, often disproportionately affected by air pollution, receive the same level of protection as more affluent areas. By reducing emissions at the source, FGD helps bridge the gap in environmental quality and promotes equity in health outcomes.

- **A Sustainable Path Forward:** In the face of mounting environmental challenges, FGD technology offers a sustainable path forward for India's energy sector. By promoting cleaner and more efficient energy production, FGD helps India transition towards a low-carbon future while simultaneously safeguarding human health and environmental integrity.

In conclusion, the importance of Flue Gas Desulfurization (FGD) in protecting human health cannot be overstated. By capturing and neutralizing harmful pollutants from coal-fired power plants, FGD technology serves as a powerful ally in the fight against air pollution, offering hope for a healthier, more sustainable future for all.

Overview of the engineering industry in India

1. **Growing Importance of Engineering:** India's engineering industry plays a pivotal role in the country's economic development, contributing significantly to GDP growth, infrastructure development, and technological advancement. The sector encompasses a wide range of disciplines, including civil, mechanical, electrical, and chemical engineering, among others.

2. **Diverse Engineering Sub-Sectors**: The engineering industry in India is diverse, covering various sub-sectors such as construction, manufacturing, energy, transportation, and telecommunications. Each sub-sector has its unique characteristics, challenges, and opportunities, contributing to the overall growth and development of the engineering landscape.

3. **Emerging Technologies:** India's engineering industry is witnessing rapid technological advancements, driven by innovation and digitalization. Emerging technologies such as artificial intelligence, Internet of Things (IoT), renewable energy, and smart infrastructure are transforming traditional engineering practices and shaping the future of the industry.

4. **Global Competitiveness:** With a large pool of skilled engineers and a favorable business environment, India's engineering industry is increasingly competitive on the global stage. Indian engineering firms are expanding their footprint internationally, undertaking projects across various countries and regions, further enhancing India's reputation as a hub for engineering excellence.

Key Sectors and Projects

- **Infrastructure Development**: Infrastructure development is a key focus area for the Indian engineering industry, with significant investments being made in roads, railways, airports, ports, and urban infrastructure. Major projects such as the construction of metro rail networks, expressways, and smart cities are driving demand for engineering services and expertise.

- **Power Generation and Transmission**: The power sector is another important sector for engineering in India, with a growing emphasis on clean energy sources and environmental sustainability. Projects related to the installation of renewable energy capacity, modernization of thermal power plants, and expansion of transmission networks are key areas of focus

- **Environmental Protection**: Given the increasing concerns about environmental pollution and climate change, projects aimed at environmental protection and sustainability are gaining importance in India's engineering landscape. Flue Gas Desulfurization (FGD) projects, in particular, are being implemented to reduce emissions of sulfur dioxide (SO_2) from coal-based thermal power plants, thus mitigating their adverse environmental impact.

- **Water Management**: With water scarcity becoming a pressing issue in many parts of India, projects related to water management, including the construction of dams, reservoirs, irrigation systems, and water treatment plants, are essential for ensuring water security and sustainable development.

Impact of FGD on Human Health and Population

1. **Hazardous Effects of SO2 and NOx Emissions:** Flue gas emissions from coal-based thermal power plants contain harmful pollutants such as sulfur dioxide (SO_2) and nitrogen oxides (NOx), which pose serious health risks to nearby communities. Exposure to these pollutants can lead to respiratory problems, cardiovascular diseases, and other adverse health effects, particularly among vulnerable populations such as children, the elderly, and individuals with pre-existing health conditions.

2. **Population Affected by FGD Emissions:** The hazardous impact of SO2 and NOx emissions from coal-based thermal power plants can affect a significant portion of the population residing in the vicinity of these plants. Depending on factors such as wind direction, proximity to the plant, and population density, the impact may extend to neighboring communities and even urban areas located several kilometers away from the source of emissions.

3. **Need for FGD Implementation:** Recognizing the adverse health effects of SO2 and NOx emissions, there is a pressing need for the implementation of Flue Gas Desulfurization (FGD) technologies in coal-based thermal power plants. FGD systems help to remove sulfur dioxide from flue gases before they are released into the atmosphere, thereby reducing air pollution and safeguarding public health.

4. **Government Initiatives**: The Indian government has taken several initiatives to address air pollution from coal-based thermal power plants, including mandating the installation of FGD systems in existing and upcoming plants. These regulatory measures aim to minimize the environmental impact of power generation activities and protect the health and well-being of the population residing in the vicinity of these plants.

To summarize, comprehending the function of Flue Gas Desulfurization (FGD) in the Indian engineering context entails appreciating its importance in reducing air pollution resulting from thermal power plants that run on coal and preserving public health. In India, where clean energy and environmental sustainability are top priorities, FGD projects will be essential to cutting emissions and guaranteeing a better, more sustainable future for all.

Role of a Project Director for various FGD projects

- Responsibilities and duties
- Skills required for effective project management
- Risk Management & Financial Management
  - Identifying project risks
  - Assessing and prioritizing risks
  - Mitigation strategies and contingency planning
  - Budgeting and cost estimation
  - Monitoring project expenses
  - Managing cash flow

Story of Project Director's Tale: Navigating the Storm

A seasoned project director by the name of Vikram formerly resided in the center of busy Mumbai. Vikram had managed Flue Gas Desulfurization (FGD) projects for years and had encountered his share of difficulties. But nothing could have equipped him for the tempest that was about to consume his most recent project.

The future seemed bright as Vikram took on the responsibility of managing a crucial FGD project. In order to complete the project on schedule and under budget, Vikram and his team set off with careful preparation and unrelenting dedication. They had no idea that unexpected challenges that could have halted their development were waiting around the corner.

Everything started with an abrupt and unanticipated change in course: the government implemented trade restrictions that prevented communication applications with China, their main technological ally, from functioning. Crucial lines of communication were cut over night, leaving Vikram and his group adrift in an uncharted area. Panic came in since they had no way of communicating with their counterparts in China.

To make matters worse, Vikram was informed with unnerving accuracy that their email server—hosted on a platform headquartered in China—had been disabled, so preventing access to vital project data and conversations. Anxious attempts to recover backups were unsuccessful since sensitive data could not be sent outside of India due to strict data protection rules.

As Vikram came to terms with the circumstance, he struggled with feelings of irritation and powerlessness. Once full of promise, the enterprise now teetered on the verge of catastrophe, trapped by forces without his control. But even in the face of confusion and uncertainty, Vikram resisted giving up.

Vikram inspired his team to persevere in the face of difficulty by drawing on his years of expertise and natural fortitude. They collaborated to come up with alternate ideas, looking for ways to save the project and lessen the effects of the communication barrier. They used their imaginations and ingenuity to create temporary lines of communication, using encrypted platforms and local servers to keep vital conversations going with their colleagues.

Concurrently, Vikram led the initiative to localize important project data, working with IT specialists and legal experts to set up safe data storage facilities in India. Even in the face of difficulties, Vikram never wavered in his will to complete the project and refused to allow outside factors determine its outcome.

Vikram and his group persisted, weathering the storm with fortitude and will as the weeks stretched into months. Together, they surmounted obstacles that seemed impossible, overcoming distance boundaries and administrative roadblocks to ensure the project's continued success.

At last, as the clouds of doubt started to clear, Vikram stood tall amid the debris, a symbol of strength and resiliency. The project turned out well in spite of the obstacles and struggles encountered along the road, which is a credit to Vikram and his team's unrelenting attitude.

Ultimately, Vikram's narrative and his FGD project served as a moving reminder of the strength of human resourcefulness and tenacity in the face of difficulty. Despite the difficulties along the way, it was precisely because of adversity that they developed resilient relationships and came out on top. With rekindled optimism and steadfast resolve, they gazed towards the horizon, prepared to take on whatever obstacles that lied ahead.

Overcoming Obstacles: The Function of a Project Director

When it comes to Flue Gas Desulfurization (FGD) projects, the Project Director plays a crucial role in guiding the project through obstacles and uncertainties in order to provide favorable results. You've gone from being a 31-year-old rookie to a seasoned leader in the industry, and your path has been shaped by priceless experiences and hard-won lessons. Let's examine the many duties and necessary competencies that characterize your position as a Project Director in different FGD projects:

Duties & Responsibilities:

1. **Overall Project Oversight:** Your main duty as the Project Director is to steer and lead strategically for every facet of the FGD projects. This entails managing the planning, carrying out, monitoring, and finishing of projects while making sure they are in line with the aims and objectives of the company.

2. **Stakeholder Management**: You will be dealing with a range of stakeholders, such as internal project teams, suppliers, contractors, government agencies, technology partners, and customers. Throughout the project lifespan, cultivating cooperation and preserving stakeholder satisfaction depend heavily on effective communication and relationship-building.

3. **Risk Management** : You are essential in recognizing, evaluating, and reducing project risks because of the dynamic nature of FGD projects and outside variables like geopolitical conflicts and worldwide pandemics. This include creating crisis response plans, contingency plans, and risk management measures to reduce downtime and guarantee project resilience.

4. **Resource Allocation**: Effective resource allocation is essential to project success because of set financial budgets and constrained deadlines. You are in charge of allocating financial, human, and material resources as efficiently as possible to complete the project on schedule and within the allocated budget.

5. **Quality Assurance and Compliance**: In FGD projects, it is crucial to uphold quality standards and adhere to contract specifications and legal requirements. You supervise the execution of quality control systems, inspection procedures, and legal safeguards to guarantee that project outputs fulfill or surpass requirements.

6. **Project Performance Monitoring**: You follow the progress of the project, spot deviations from the plans, and take corrective action as necessary to maintain it on schedule by using data-driven metrics and performance indicators. Transparency and accountability in project management are facilitated by regular performance evaluations and reporting systems.

Competencies Needed for Successful Project Management:

- **Leadership**: As a project director, you must possess excellent leadership abilities in order to encourage teamwork, inspire and motivate project teams, and face obstacles head-on with resiliency and resolve.

- **Strategic Thinking**: In the fast-paced world of FGD projects, strategic thinking helps you foresee future trends, spot possibilities, and develop proactive plans to deal with new problems.

- **Communication**: Clear and concise communication is crucial for setting project objectives, communicating expectations, settling disputes, and keeping lines of communication open with all stakeholders.

- **Adaptability**: In the face of unanticipated obstacles, your capacity to adjust to shifting conditions and handle uncertainty—as demonstrated by your involvement in international incidents like as the Covid-19 outbreak and geopolitical tensions—is essential to the success of any project.

- **Problem-Solving**: In order to assess complicated problems, come up with original ideas, and make wise judgments under pressure, a project director has to be proficient in problem-solving techniques. This is especially true in scenarios where supply chain interruptions or outside influences affect the project's completion.

- **Team Management**: Managing a team effectively entails creating a cooperative work atmosphere, giving team members autonomy, offering mentorship and direction, and utilizing their varied backgrounds and specialties to accomplish project goals as a group.

Other Responsibilities of the Project Director in Risk Management and Financial Management for FGD Projects

As the Project Director in charge of Flue Gas Desulfurization (FGD) projects, my duty is to guarantee careful focus on risk management and financial supervision. From the standpoint of a project director, I will outline my approach to three crucial aspects:

- **Identifying Project Risks - Meticulous Risk Assessment:** Performing a thorough examination to identify possible risks that are unique to FGD projects, such as adherence to regulations, technical obstacles, and interruptions in the supply chain.

- **Stakeholder Engagement:** Engaging with project team members, suppliers, and regulatory agencies to collect insights and opinions about possible risks and uncertainties.

- **Evaluating and Ranking Risks - Risk Quantification:** Evaluating the probability and consequences of identified risks in order to prioritize them according to their importance to project goals and results.

Mitigation Strategies and Contingency Planning:

- **Creating Mitigation Plans:** Creating strong strategies to reduce the impact of identified hazards, which include preventative measures and proactive reaction plans.
- **Contingency Planning:** Creating backup plans and setting aside resources to minimize the consequences of unexpected events or hazards that may arise during project implementation.

Budgeting and Cost Estimation: -

- **Thorough Budgeting:** Creating comprehensive budgets that include all project expenditures, such as equipment acquisition, construction charges, labor costs, and contingency funds.
- **Precise Cost estimate:** Utilizing dependable cost estimate methods, such as bottom-up estimation and parametric modeling, to properly predict project costs.

Tracking Project Expenditures:

- **Real-Time Expense Tracking:** Establishing mechanisms and procedures to monitor project expenditures in real-time, ensuring compliance with budgetary limitations and rapidly detecting any violations.
- **Variance Analysis:** Performing variance analysis to comparing the actual spending of a project with the planned costs and identifying areas that need corrective action or change.

Cash Flow Management: -

- **Cash Flow Projection**: Anticipating the cash flow needs at different stages of the project to ensure sufficient funds are available to pay financial commitments and cover project costs.
- **Enhancing Working Capital Efficiency:** Enforcing efficient management strategies to maximize cash flow and mitigate the possibility of liquidity shortages or project financing delays.

As Project Director, my primary focus is on effectively overseeing risk management and financial aspects to guarantee the successful and efficient completion of FGD projects. Through conscientiously addressing possible risks, maintaining stringent financial discipline, and implementing effective mitigation techniques, my aim is to protect project results and provide value to stakeholders.

In conclusion, your experience leading many FGD projects as a Project Director demonstrates the fortitude, flexibility, and leadership needed to overcome obstacles and produce positive results in challenging and ever-changing settings. Your stories demonstrate the value of strategic leadership and proficient project management in achieving project success in the face of challenges.

Tender Participation Stage

- Introduction to tendering process in India
- Preparing for tender submissions
- Strategies for successful tender participation

Effective Strategies for Success in FGD Projects: Navigating the Tender Participation Stage

Opening the Business Opportunity in India :

The tender participation stage in Flue Gas Desulfurization (FGD) projects is a crucial step that requires careful planning and strategic decision-making in order to secure profitable contracts. Within the framework of our obligations as described in the FGD project scenario, this phase entails forming connections with prospective customers, comprehending market dynamics, and creating competitive proposals that showcase our proficiency and capacities. Let's examine the complexities of participating in tenders in the FGD industry and discuss techniques for achieving success.

Analyzing the Framework of the Tendering Process in India:

1. **Regulatory Framework**: - The bidding process in India is regulated by several entities, such as government agencies, public-sector firms, and private organizations. Comprehending the legislation, procurement procedures, and assessment criteria established by these organizations is essential for adhering to the rules and actively engaging in bidding processes.

2. **Market Dynamics:** - The FGD industry in India is characterized by constant change and intense competition, driven by an increasing need for solutions that ensure environmental compliance. - Engaging in market research to comprehend industry trends, upcoming technologies, and competitor tactics is crucial for making well-informed decisions while participating in tenders.

3. **Client ties:** - Establishing robust ties with prospective clients is a fundamental aspect of achieving success in tender engagement. Engaging in networking activities with procurement departments, participating in industry events, and using established relationships aid in the identification of tender chances and comprehension of client needs.

Step 1 : Preparing for Tender Submissions:

Technical Proficiency:

- Displaying technical proficiency and experience in FGD projects is essential for building trust with customers.
- By stressing our track record of successful former projects, displaying our cooperation with prominent international partners, and highlighting our expertise in FGD technologies, we may boost our bid bids.

Financial Viability:

- Guaranteeing financial sustainability is of utmost importance when formulating competitive proposals that strike a balance between profitability and affordability.
- The user did not provide any text. Performing comprehensive cost evaluations, which include the acquisition of equipment, its installation, and ongoing operating costs, aids in establishing an accurate bid price that is in line with the project's goals.

Risk Management:

- The process of recognizing and reducing potential project hazards is crucial for building trust in customers and guaranteeing project triumph.
- The user did not provide any text. Performing thorough risk evaluations, formulating strong risk reduction plans, and offering customers reassurances about our risk management skills bolster the credibility of our proposals.

Step 2 : Effective Approaches for Achieving Successful Involvement in Tender Processes:

Competitive Pricing:

- It is crucial to provide competitive pricing in order to get contracts in the FGD industry and ensure profitability.
- Employing a methodical pricing strategy, based on thorough market study and cost evaluations, enables us to position our bids advantageously in comparison to our rivals.

Ensuring openness and honesty

- Throughout the bidding process is essential for establishing confidence with customers. The user did not provide any comments on our Techno commercial offer.
- Our dedication to ethical business practices is shown via the provision of precise and thorough bid paperwork, strict adherence to bidding dates, and guaranteeing compliance with regulatory standards.

Continuous Improvement:

- Adopting a culture of continuous improvement is crucial for refining tendering methods and boosting competitiveness.
- The user did not provide any text. Seeking input from previous tender experiences, evaluating bid results, and integrating acquired knowledge into future bidding procedures assist to continuous development and achievement.

Strategies for successful tender participation

To summarize strategies for achieving success in participating in tenders in the FGD sector requires a comprehensive strategy that includes ensuring compliance with regulations, doing thorough market research, establishing strong relationships, engaging in rigorous preparation, and upholding ethical standards. Through the use of technical proficiency, financial experience, and a steadfast dedication to achieving the highest standards, we establish ourselves as the choice collaborators for customers in need of dependable resolutions for their FGD projects. By using strategic planning, risk management, and continuous development, we confidently traverse the tender participation stage, ensuring the acquisition of contracts and driving success in the dynamic FGD market.

Project Initiation Phase
- Project kick-off and team formation
- Establishing project goals and objectives
- Initial planning and scheduling

Project Initiation Phase: Understanding the Origins of FGD Projects in India

Within the domain of Flue Gas Desulfurization (FGD) projects in India, the Project Initiation Phase acts as the origin, establishing the foundation for a successful undertaking. This stage entails coordinating the initiation of the project, assembling a unified team, and carefully defining project goals and objectives. Furthermore, the start step includes a thorough assessment of contract provisions to proactively address any disputes in the future. Now, let's explore the complexities of each individual element:

1. Initiation of Project and Establishment of Team:

- **Involving Stakeholders:** Initiating a FGD project involves assembling a heterogeneous group of stakeholders, such as project managers, engineers, environmental experts, and procurement specialists. A kick-off meeting functions as a forum for synchronizing team members with the project's goal, promoting cooperation, and delineating their individual roles and obligations.

- **Developing a Unified Team:** The outcome of a FGD project depends on the efficiency of the team. Project leaders are responsible for organizing team-building exercises, promoting transparent communication, and fostering a culture of collaboration. Within a multidisciplinary setting, the proficiency of every team member plays a crucial role in determining the overall success of the project.

2. Defining Project Goals and Objectives:

- **Establishing explicit goals:** The Project Initiation Phase is an ideal opportunity to establish unambiguous and quantifiable project goals. Within the framework of FGD projects in India, these goals may include ensuring environmental compliance, achieving emission reduction targets, adhering to regulatory requirements, and completing the project within the specified timeframe.

- **Meeting Stakeholder Expectations:** Establishing project objectives entails ensuring their congruence with the expectations of many stakeholders, such as clients, regulatory organizations, and environmental agencies. Comprehending and integrating these expectations into project goals guarantees a shared comprehension of success criteria.

Story related to The Misadventures of Project Manager Mishra

In the city of Mumbai, there resided a project manager called Mishra. Mishra was renowned for his unwavering commitment and meticulousness, however he had a talent for sometimes misconstruing project objectives in a delightfully humorous manner.

On a clear morning, Mishra was given a new project task: to supervise the construction of a tall office building in the center of the city. Mishra eagerly embraced the chance to demonstrate his exceptional management skills and enthusiastically embarked on the project, equipped with detailed plans and an unwavering passion.

While examining the project scope paper, Mishra's eyes expanded with eager expectation. The text said, "Erect a structure that is of great height and evokes a sense of wonder." Filled with grand aspirations of towering structures, Mishra enthusiastically launched on his assignment.

Unknown to Mishra, his understanding of "stands tall and inspires awe" was significantly distorted. Mishra's approach diverged from emphasizing architectural grandeur and structural integrity, as he instead sought to infuse the structure with a playful and unconventional atmosphere.

With this newly acquired inspiration, Mishra directed the construction team to include distinctive architectural aspects throughout the structure. The outside of the building was embellished with colossal sculptures depicting elephants in motion, while the rooftop garden was decked with vibrant neon lights and disco balls. Mishra strongly advocated for the installation of a roller coaster that descended in a spiral fashion from the highest floor to the lowest level, greatly perplexing the building crew.

During the course of the project, speculation started circulating about Mishra's unorthodox methodology in architectural design. Onlookers would pause and gaze in astonishment at the remarkable scene happening in front of them. The local media used the term "The Mishra Marvel" to highlight its unconventional nature and boldness.

Nevertheless, Mishra remained completely oblivious to the havoc he had initiated. He saw each curve and loop of the roller coaster, as well as every imaginative sculpture, as a brilliant display of ingenuity, serving as evidence of his visionary guidance.

Mishra's misunderstanding only became evident after the project was almost finished. The customer, who had desired a streamlined and contemporary office building, was horrified by the appearance that lay before them. "This is not what we agreed to!" they cried, their expressions filled with shock and horror as they saw the brightly illuminated elephants and roller coaster attractions.

Mishra made a desperate effort to rescue the situation by attempting to explain his view of the project scope. Nevertheless, his elucidations further exacerbated the client's confusion, and before long, utter disorder prevailed on the building site.

Ultimately, Mishra gained a significant insight into the significance of effective communication and comprehensive comprehension of project scopes. Having seen the consequences of his mistakes, Mishra, also known as "The Mishra Marvel," made a solemn promise to never underestimate the potency of exact language and a well defined project scope in the future. Despite the roller coaster being disassembled and the neon lights losing their brightness, Mishra's reputation endured as a cautionary story in the records of project management mythology.

3. Preliminary Planning and Scheduling:

- **Thorough Project Planning:** Effective project planning is essential for achieving success. In the FGD industry, strict adherence to deadlines is crucial. Therefore, a comprehensive project plan that includes the scope, resources, timeframe, and budget is very necessary. The planning phase encompasses the evaluation of potential risks and the development of solutions to minimize their impact.

- **Establishing Feasible Timetables**: Due to the ever-changing nature of FGD projects, it is crucial to establish accurate and achievable timelines. During the early planning phase, one must take into account many possible obstacles, including obtaining governmental permits, procuring necessary technologies, and adhering to building timetables. An accurate timetable functions as a guide for the next stages of the project.

4. Assessment of Contract Terms:

- **Comprehensive Contract Examination:** The start step is an ideal opportunity to conduct a thorough assessment of contract conditions. When doing FGD projects in India, it is crucial to conduct a comprehensive examination of contract conditions due to the potential influence of external variables such as legislative changes and geopolitical events on project dynamics. This includes payment schedules, performance guarantees, and systems for resolving disputes.

- **Strategies for Avoiding Conflict:** It is essential to identify possible causes of dispute in contract conditions and take proactive measures to remedy them. Utilizing precise and unequivocal language, including dispute resolution provisions, and establishing a shared knowledge of expectations are effective strategies for preventing problems during project implementation.

The Project Initiation Phase in FGD projects in India establishes the foundation for a process that requires careful planning, efficient team building, and a comprehensive assessment of contractual conditions. Project leaders may ensure a smoother and more effective project execution by aligning project objectives with stakeholder expectations, promoting team cohesiveness, and resolving any problems in contract terms. This phase is crucial for the successful implementation of the FGD project in India. It will provide a strong foundation, ensuring that the project meets environmental regulations, satisfies stakeholders, and adheres to high project management standards.

Project Execution

- Managing resources effectively
- Communication and coordination with stakeholders
- Monitoring progress and making adjustments

Achieving Proficiency in Project Implementation in the Ever-Changing Realm of FGD Projects

In the rapidly changing and constantly developing field of Flue Gas Desulfurization (FGD) projects, achieving success in project execution is not simply a matter of adhering to a pre-established plan. It is a dynamic undertaking that necessitates flexibility, versatility, and a profound comprehension of the distinctive difficulties inherent in the industry. Project managers must adeptly traverse a complex array of factors, including shifting FGD layouts, increasing end-customer needs, and changeable government legislation, in order to assure the success of a project. This guide will provide a detailed examination of project execution in the FGD sector, focusing on tactics for efficient resource management, effective stakeholder engagement, and progress monitoring for timely modifications.

Efficient Resource Management

- **Allocation of Resources in a Dynamic Manner**
  Within this part, we will examine the difficulties associated with the management of resources in the specific context of FGD projects. We will explore the significance of predicting fluctuating resource requirements, constructing backup strategies, and harnessing cross-functional teams to enhance efficiency and production.

- **Flexibility in Resource Planning**
  In this phase, we will explore the need of adopting an adaptable strategy for resource allocation in response to changing project demands. We will explore techniques for adjusting resource allocation tactics, accessing alternate supply sources, and minimizing the effects of unforeseen changes on project schedules and budgets.

The narrative starts with a prominent engineering company assigned with the responsibility of executing a Flue Gas Desulfurization (FGD) project for a coal-fired power station located in India. The project was advancing in accordance with the established plan, with resources assigned in accordance with the original project scope and needs. Nevertheless, as the project progressed, unanticipated alterations started to arise.

Obstacles Encountered:

1. **Modifying Project Scope:** The customer requested changes to the FGD layout during the middle of the project, which required alterations to the strategy for allocating resources.

2. **Supply chain disruptions**: Global interruptions in the supply chain resulted in delays in the delivery of essential equipment, which had an effect on project timetables.

3. **Budgetary limitations:** The project team encountered financial restrictions, necessitating them to discover economical resolutions while upholding project excellence.

Flexible Resource Allocation:

To address these problems, the project team acknowledged the need of including adaptability into resource planning. To mitigate the effects of unexpected changes, they used the following measures to adapt resource allocation tactics:

- **Optimizing Resource Allocation Strategies:**
  - o The project team did a comprehensive reevaluation of the resource needs in response to the modifications made to the project's scope.
  - o They redistributed human resources and equipment to the project's most crucial sectors, guaranteeing that essential activities were given priority.
  - o Specialized teams were created to tackle particular project obstacles, using the knowledge and skills of individuals from several departments to discover inventive resolutions.

- **Utilizing Alternative Supply Sources:**
  - o To ensure the project's progress remains uninterrupted despite delays in the supply chain, the project team investigated other options for procuring equipment.
  - o They formed alliances with nearby suppliers to accelerate the distribution of crucial resources, reducing dependence on overseas suppliers impacted by logistical obstacles.

- **Mitigating the Impact of Unexpected Alterations:**
  - o In order to deal with financial limitations, the project team used cost-effective strategies while maintaining the high standards of the project.
  - o They enhanced resource allocation, simplified operations, and secured advantageous agreements with suppliers to cut costs while enhancing benefits.

Result: The project team effectively managed the difficulties caused by shifting project requirements due to their adaptable resource planning methodology. Despite encountering challenges, they successfully adjusted promptly, sustained project progress, and completed the FGD project punctually and within the allocated budget. The client was pleased with the team's dexterity and dedication to problem-solving, solidifying the firm's image as a reliable partner in FGD project management.

Communicating and Coordinating with Stakeholders

- **Clear and Honest Communication**
  This section will analyze the crucial significance of transparent and unobstructed communication in the implementation of a project. We will explore the significance of consistently engaging with stakeholders, delivering timely communications, and cultivating a cooperative environment to guarantee that expectations are aligned and new concerns are addressed proactively.

- **Engaging Stakeholders in an Agile Manner**
  In this discussion, we will examine the need of using agile stakeholder engagement tactics to effectively address the changing dynamics of a project. We will examine the significance of regularly holding progress meetings, seeking input, and adjusting communication methods to cater to the requirements of various stakeholders.

Monitoring Progress and Making Adjustments

Monitoring Progress in Real-Time
This phase of project will explore the significance of monitoring progress in real-time to guarantee the success of a project. We will explore the use of sophisticated project management tools and methodologies to track essential performance metrics, detect obstacles, and promptly resolve any arising problems.

- **Utilizing Advanced Project Management Tools:**
  - **Digital Collaboration Platforms:** Leveraging cloud-based collaboration tools facilitates real-time communication and information sharing among project teams spread across diverse geographical locations.
- **Project Management Software:** Implementing sophisticated project management software enables the tracking of project timelines, resource utilization, and critical milestones in real-time.

Agile Decision-Making
We will examine the fundamental concepts of agile decision-making in FGD projects. We will explore techniques for acquiring pertinent data, evaluating the consequences of suggested modifications, and promptly making well-informed choices to maintain project progress while considering financial limitations and price swings.

Within this concluding piece, we will succinctly outline the essential points derived from our investigation into project implementation within the ever-changing realm of FGD projects. We will emphasize the significance of agility, flexibility, and efficient communication in successfully navigating the difficulties associated with resource management, stakeholder involvement, and progress tracking. Ultimately, we will highlight the significance of ongoing enhancement in promoting project triumph and guaranteeing the timely and cost-effective delivery of FGD projects, while meeting the expectations of all parties involved.

This thorough reference provides project managers in the FGD sector with in-depth analysis of every area of project execution. It gives them with the necessary information and methods to effectively traverse the complexity of their business. By efficiently allocating resources, maintaining clear and open communication, and making quick and adaptable decisions, project managers may successfully navigate the obstacles presented by changing project circumstances and achieve outcomes that satisfy both clients and stakeholders.

Site Management

- Setting up and organizing construction sites
- Ensuring safety regulations compliance
- Handling site-specific challenges

Site management in flue gas desulfurization (FGD) projects: Addressing challenges in the Indian context.

Managing sites in Flue Gas Desulfurization (FGD) projects in India has distinct problems, including tasks such as site organization and adherence to safety rules. One major challenge is the storage and safeguarding of items provided at client premises. This chapter will explore the intricacies of site management in FGD projects in India, with a special emphasis on addressing obstacles pertaining to material storage, adherence to safety regulations, and managing site-specific concerns.

The Story of the Misplaced Material: A Time-Constrained Race in FGD Site Commissioning

In the middle of a busy construction site in India, where a Flue Gas Desulfurization (FGD) project was taking place, a sequence of events occurred that challenged the project team's resilience and inventiveness to the highest degree.

In Uttar Pradesh, a group of engineers and technicians had been diligently working on the installation and commissioning of Flue Gas Desulfurization (FGD) equipment at a thermal power plant. The project had been advancing seamlessly, with careful strategizing and synchronization guaranteeing the punctual achievement of each milestone. Nevertheless, destiny had an unexpected twist for them.

As the day of commissioning neared, a palpable sense of excitement permeated the atmosphere. The final stage of the project was set to be seen by representatives from the customer, government authorities, and stakeholders. This event marked the significant achievement of months of diligent effort and commitment.

However, at the precise moment when all the elements seemed to be aligning perfectly, a catastrophic event occurred. During a regular inventory audit, it was found that a vital component of the FGD equipment was missing. The project crew was filled with panic as they comprehended the seriousness of the problem. Deprived of this crucial component, the process of commissioning the equipment would become unfeasible, posing a risk to the whole project's schedule and reputation.

Under the pressure of time and the scrutiny of all stakeholders, the project team quickly started their actions. Telephone conversations were initiated, urgent gatherings were organized, and other strategies were quickly devised. The task of finding the lost material and ensuring the commissioning could continue as scheduled was a race against time.

As time progressed, the level of stress increased significantly. Each minute that passed resulted in the postponement of a crucial project milestone and the possibility of negative financial consequences. However, in the middle of the disorder, the project team maintained their firm resolve, driven by a collective will to surmount the impediment impeding their progress.

Ultimately, via a fortuitous turn of events and unwavering determination, the misplaced material was discovered at a different project location situated several hundred kilometers apart. Immediately, preparations were undertaken to swiftly transfer the material to the commissioning location.

At dusk on the day before the commissioning day, the project crew felt a sense of relief when the previously absent material arrived at the site, just in time. The equipment was efficiently put into operation and the project was effectively finished, thanks to the team's resolute dedication and exceptional collaboration.

Following the traumatic experience, the project team not only emerged triumphant but also shown increased resilience and fortitude. The narrative of the misplaced substance served as a monument to their fortitude, flexibility, and steadfast dedication to achieving outcomes under challenging circumstances. While celebrating their well-earned triumph, they were aware that no obstacle was insurmountable when they stood unified as a cohesive team.

Establishing and Arranging Construction Sites:

Challenges in Material Storage:

- **Delineation of Customer Scope and Supplier Responsibility**:
  While the customer scope outlines the duties for material storage, suppliers often encounter difficulties arising from insufficient storage facilities at customer locations.

- **Risk of Material flaws**: Inadequate storage conditions, such as exposure to severe weather or insufficient protection, heighten the likelihood of material flaws, resulting in delays and increased expenses.

Timely Erection and Commissioning: -

- **Unanticipated Delays:** The need for departmental permissions and bureaucratic procedures may frequently result in unanticipated delays in the erection and commissioning operations, which can have an impact on the project timeframes.
- **Learning Curve for Foreign OEMs:** Foreign original equipment manufacturers (OEMs) face a period of adjustment in understanding and adapting to the regulatory procedures and special demands of Indian customers. This might result in delays in effectively resolving difficulties specific to each site.

Ensuring Compliance with Safety Regulations:

Regulatory Compliance Challenges: -

- **Complex Regulatory environment:** Successfully navigating the intricate regulatory environment in India necessitates a comprehensive comprehension of safety rules and compliance prerequisites.
- **Diverse Interpretation**: The understanding of safety standards might differ across different areas and local governing bodies, requiring continuous attentiveness and communication to guarantee adherence.

Mitigating Safety Risks: -

- **Preventive Safety Measures:** Enforcing preemptive safety measures, such as periodic safety audits, training initiatives, and safety regulations, mitigates the likelihood of accidents and guarantees a secure working environment for all parties involved.

Addressing Site-Specific Difficulties:

- **Customer-Supplier Collaboration**: -
    - o **Efficient Communication:** Establishing efficient communication channels between customers and suppliers is essential for addressing site-specific obstacles and resolving issues immediately.
    - o **Collaborative Problem-Solving**: Promoting collaborative problem-solving methods cultivates a feeling of cooperation and guarantees a shared comprehension of project specifications and limitations.

- **Risk Management Strategies:** -
    - o **Risk Identification and Mitigation:** Implementing proactive measures to identify and mitigate risks specific to the project site, such as addressing challenges related to material storage and approval delays, necessitates the use of comprehensive risk management strategies that are customized to meet the unique requirements of each site.
    - o **Contingency Planning:** The process of creating backup plans for possible interruptions, such as the presence of faulty materials or unexpected delays, aids in reducing the negative effects on project schedules and budgets.

Efficient management of sites in Flue Gas Desulfurization (FGD) projects in India requires a comprehensive strategy to address obstacles pertaining to material storage, adherence to safety regulations, and site-specific concerns. Through proactive measures and the implementation of efficient risk management procedures, project managers can guarantee the successful implementation of FGD projects in the ever-changing and demanding Indian environment. Effective collaboration between customers and suppliers, strict attention to safety rules, and the ability to adapt to site-specific needs are crucial components in successfully managing sites and ensuring project success in the FGD industry in India.

Quality Assurance and Control

- Implementing quality management systems
- Conducting inspections and audits
- Ensuring adherence to standards and specifications

Guaranteeing Superior Quality in FGD Site Implementation: The Vital Role of Contract Management

Quality assurance and control are essential for the successful implementation of FGD site projects. To maintain the highest levels of quality throughout the project lifespan, project stakeholders may do this by creating strong quality management systems, performing frequent inspections and audits, and guaranteeing strict compliance with standards and specifications. By doing this, they not only reduce risks and save expensive rework, but also preserve their dedication to providing excellent results that match or beyond customer expectations. Quality assurance and control play a crucial role in contract management for FGD site execution, ensuring that every part of the project complies with the highest standards of quality and integrity.

Story related to Quality – The Dinner Party

Amidst a thriving Indian kingdom, a well-known contractor was given the responsibility of organizing a grand dinner party to commemorate a significant event in the royal family's history. The contractor, renowned for their expertise in orchestrating extravagant gatherings, enthusiastically embraced the opportunity, pledging a banquet befitting nobility.

Prior to the occasion, the contractor made thorough preparations. The dinner space was converted into a magnificent display, adorned with vivid decorations, majestic furnishings, and polished kitchenware. The scent of alluring spices permeated the atmosphere as proficient cooks crafted a lavish assortment of meals that showcased the varied and opulent nature of Indian cuisine.

As the evening progressed, visitors, dressed in traditional clothing, gathered to participate in the celebrations. The hall reverberated with laughter and pleasure as the party began. The contractor, who was pleased with their thorough preparations, observed as the royal family and notable visitors fully engaged in the festivities.

Nevertheless, destiny took an unforeseen twist. During the joyful event, a perceptive member of the royal family saw an unpleasant visitor—a little fly hidden discreetly in one of the dishes. The disclosure created a somber atmosphere throughout the festivities, disrupting the unity and evoking unease among the attendees.

Notwithstanding the contractor's many apologies and promises of their dedication to cleanliness, the harm had already been inflicted. The monarch, who upholds customs and integrity, was very disappointed by the mistake. The contractor swiftly and decisively addressed the repercussions of their irresponsibility, resulting in a penalty that reverberated throughout the whole kingdom.

The narrative of the ill-fated dinner gathering became into a cautionary parable inside the royal court and extended to other circles. It was a powerful reminder that even the most extravagant celebrations in Indian hospitality might be spoiled by the smallest mistake. The contractor, previously renowned for their proficiency, now carried the burden of a great lesson—that in the complex choreography of cultural festivities, meticulousness and a steadfast dedication to excellence were the genuine treasures in the hierarchy.

As the tale reverberated around the marketplaces and grand residences, it transformed into a moral lesson for all those who are tasked with the duty of organizing festive events. This highlighted the vulnerability of trust and the need to maintain the utmost principles, to prevent the joy of celebration from souring due to negligence.

Learning: Quality assurance and control are crucial for the effective implementation

Quality Assurance at Flue Gas Desulfurization (FGD) projects on-site.

This phase we explore the crucial significance of quality management systems, inspections, and compliance with standards and specifications in the administration of FGD site execution contracts. Ensuring quality perfection is crucial for the success of a project, given the complex construction procedures and strict regulatory standards involved.

Enacting Quality Management Systems:

- **Strong Quality Framework**: -
  - o **Systematic Approach**: Implementing a strong quality management system offers a systematic framework for monitoring and ensuring quality at every stage of the project.

  - o **Documentation and processes:** The act of documenting quality processes, standards, and specifications guarantees uniformity and openness in implementation, facilitating efficient quality control methods.

- **Continuous Improvement: -**
  - o **Feedback Mechanisms:** The implementation of feedback mechanisms enables the ongoing assessment and enhancement of quality management procedures, cultivating a culture characterized by exceptional performance and creativity.
  - o **Training and Development:** Allocating resources towards training and development initiatives provides site staff with the necessary expertise and understanding to maintain high quality standards and successfully tackle growing obstacles.

Performing Inspections and Audits:

- **Regular Site Inspections**: -
  - o **Systematic Checks**: Performing frequent inspections at different construction phases allows for early identification of quality concerns and deviations from project specifications.
  - o **Compliance with Checklists**: By adhering to predetermined inspection checklists, one may assure a comprehensive and uniform evaluation of craftsmanship and conformity to established standards.

- **Independent Audits: -**
  - o **Objective Evaluation:** Employing independent auditors to carry out regular audits offers an impartial evaluation of quality performance and adherence to contractual obligations.
  - o **Detection of Non-Conformities:** The process of conducting audits allows for the identification of non-conformities and areas that need improvement. This enables the

quick implementation of remedial measures, which helps to mitigate risks and enhance the overall quality of the project.

Guaranteeing Compliance with Standards and Specifications:

- **Regulatory Compliance: -**
  - o **Thorough Comprehension:** It is crucial to have a thorough comprehension of regulatory requirements and industry standards in order to align project execution with legal and environmental regulations.
  - o **Proactive Compliance Measures**: By implementing proactive compliance measures, such as conducting environmental impact assessments and monitoring emissions, the risk of regulatory infractions and the resulting fines may be minimized.

- **Specifications Adherence: -**
  - o **Comprehensive Specifications Review:** Conducting meticulous evaluations of project specifications and design documentation guarantees conformity with client expectations and industry standards.
  - o **Supplier and Contractor Compliance:** Ensuring that suppliers and contractors follow the specified requirements via contractual agreements and performance monitoring procedures preserves the uniformity and reliability of project outputs.

Performance Guarantee Test

- Preparing for performance testing
- Conducting performance evaluations
- Addressing any deficiencies or issues

Achieving Optimal Performance in FGD Projects: The Performance Guarantee Test

The Performance Guarantee Test (PGT) is a crucial milestone in Flue Gas Desulfurization (FGD) projects, since it verifies that the deployed systems conform to the stated performance criteria. This chapter explores the detailed procedure of planning for and carrying out performance assessments in FGD projects, while also addressing any shortcomings or problems that may occur during testing. The PGT, or Project Greenlight Test, is crucial for ensuring both environmental compliance and operational efficiency, and serves as the final measure of project success.

Preparing for Performance Testing:

- **Thorough Test Plan through Engineering team involvement**: -
    - **Thorough Preparation:** Developing a thorough test plan that clearly defines testing processes, performance requirements, and measurement parameters enables a methodical approach to the PGT.
    - **Explicit Goals:** Establishing unambiguous goals and measurable benchmarks for the performance testing procedure offers a guide for assessing the efficiency and effectiveness of the system.
    - **Establishing Precise Objectives:** Precisely establishing the goals of the performance testing procedure is crucial for directing the testing endeavors and guaranteeing congruence with project objectives. The objectives may include the validation of system design, the verification of compliance with regulatory requirements, and the assessment of operating efficiency.
    - **Objectives for measuring success:** Defining success criteria that clearly outline the anticipated results of performance testing serves as a standard for assessing the efficiency of the FGD systems. The criterion should possess the qualities of measurability, achievability, and alignment with the expectations of project stakeholders.

- **Resource Allocation:** -
    - **Resource Identification:** The process of determining the essential resources, such as equipment, staff, and testing facilities, to fulfill all the criteria for performing precise and dependable performance assessments.
    - **Scheduling and Coordination:** Optimal scheduling and coordination of resources optimize the testing process, reducing downtime and boosting production.

Performing Performance Assessments at Site:

- **Thorough Testing processes:** -
    - **Systematic Approach:** Adhering to rigorous testing processes as outlined in the established test plan guarantees consistency and the ability to repeat performance assessments.
    - **Authentic circumstances:** Establishing test circumstances that closely replicate real-life operational situations allows for precise evaluation of system performance under ordinary operating conditions.

Performance testing is an essential stage in Flue Gas Desulfurization (FGD) projects, verifying that the implemented systems conform to the given performance criteria. In order to adequately prepare for performance testing, it is necessary to create a thorough test strategy that clearly defines the testing methodologies, performance requirements, and measurement parameters. Furthermore, it is essential to establish unambiguous goals and measurable benchmarks to serve as a guide for assessing the performance and usefulness of the system.

- **Criteria for evaluating performance:** It is crucial to establish precise performance standards that the FGD systems must adhere to throughout testing. These metrics may include factors such as the effectiveness of pollution removal, the dependability of the system, the speed of reaction, and the stability of operation.
- **Parameters for measurement:** It is essential to determine the specific measurement criteria and metrics that will be used to assess the performance of the system. This may require identifying the specific instruments and techniques for collecting data to be used throughout the testing process.

To guarantee a methodical approach to performance testing in FGD projects, project teams may methodically create a complete test strategy and establish precise goals and success criteria. This not only enables the fast execution of testing but also establishes a framework for assessing system performance and functionality based on predefined criteria and requirements.

Collection and Analysis of Data:

- **Data Integrity: -**
  - o **Ensuring Accuracy:** It is of utmost importance to ensure the accuracy and integrity of the data acquired during performance testing. This entails developing rigorous documentation and validation procedures to ensure the precision and dependability of the collected data.
  - o **Comprehensive Inspection:** Performing a comprehensive analysis of test results is crucial. This procedure entails meticulously examining the data to detect trends, anomalies, and possible opportunities for improvement. It facilitates informed decision-making and provides the basis for successful corrective actions.

Addressing Any Deficiencies or Issues:

- **Identification of Deficiencies: -**
  - o **Thorough Evaluation:** Following the testing process, it is crucial to execute a thorough evaluation in order to identify any shortcomings or performance issues that may have emerged. This stage entails a thorough examination of the testing outcomes to guarantee a full comprehension of the system's functioning.
  - o **Fundamental Cause Analysis:** Conducting a thorough root cause analysis is essential. This approach entails a thorough examination aimed at pinpointing the underlying factors that are accountable for any deficiencies in performance. In order to achieve successful resolution, it is crucial to comprehend the underlying reasons, which may be from design flaws, operational errors, or external factors.

- **Strategies for Addressing or Alleviating the Problem:**

- o **Remedial Measures:** It is crucial to promptly adopt corrective steps in Swift to address discovered shortcomings and improve system efficiency. The necessary measures to rectify the situation may include modifications to the design, improvements in operations, or further training to specifically tackle the identified problems.
- o **Ongoing Enhancement:** Fostering a culture of constant development is vital. Incorporating knowledge acquired from performance testing into future project iterations guarantees a continuous improvement and augmentation of system performance. This iterative procedure enhances the overall effectiveness and optimization of the FGD system.

To successfully negotiate the difficulties of performance testing in FGD projects, project teams should prioritize data integrity, conduct detailed evaluations, implement corrective measures, and engage in continuous improvement programs. These methods not only rectify current shortcomings but also enhance the long-term effectiveness and durability of the established solutions.

The Performance Guarantee Test is the last step of FGD projects, when precise planning, rigorous testing, and attentive analysis come together to validate the system's performance and functioning. To guarantee that FGD systems achieve or surpass stated performance criteria, project stakeholders should follow extensive test plans, conduct thorough performance reviews, and fix any faults or problems that may surface during testing. By doing this, they not only ensure adherence to environmental regulations and improve operational effectiveness, but also showcase their dedication to achieving excellence in all aspects of executing FGD projects.

Project Closure

- Completing project deliverables
- Conducting performance evaluations
- Lessons learned and continuous improvement

A Thorough Approach to Project Closure

As a Project Director responsible for managing Flue Gas Desulfurization (FGD) projects, the closing phase is a crucial stage that requires careful and detailed attention.

Story related to Project Closure

Once the project closure at Uttar Pradesh, there was a team of dedicated engineers and project managers working tirelessly on a monumental Flue Gas Desulfurization (FGD) project. For months, they had poured their sweat and expertise into every aspect of the project, from planning and procurement to construction and commissioning.

As the project neared completion, there was a palpable sense of anticipation in the air. The team had overcome numerous challenges and obstacles along the way, but their perseverance had paid off, and the FGD system was now ready for operation. However, amidst the excitement of reaching the finish line, there was also a bittersweet realization that the time had come to bid farewell to the project that had consumed their days and nights for so long.

With the final tests successfully completed and the client's approval obtained, it was time for the official closure of the project. The project director gathered the team for a final meeting to reflect on their journey together and celebrate their achievements. There were smiles and laughter as they reminisced about the challenges they had overcome and the milestones they had reached.

But amidst the celebration, there was also a sense of solemnity as the team acknowledged that their time together on this project was coming to an end. Each member had poured their heart and soul into the project, forging bonds that would last a lifetime. And now, as they prepared to part ways, there was a sense of gratitude for the opportunity to work together and a shared pride in what they had accomplished.

As the final paperwork was signed and the last details were attended to, the team gathered one last time for a farewell dinner. There were heartfelt speeches and words of appreciation as they raised their glasses to toast to the success of the project and the bonds of friendship that had been forged along the way.

And as they said their final goodbyes and went their separate ways, there was a sense of satisfaction knowing that they had left behind a legacy that would endure for years to come. For in the world of project management, closure is not just the end of one chapter but the beginning of another, filled with new challenges, new opportunities, and new adventures to be had.

Here is my systematic approach to the essential components of project closure, aimed at achieving a smooth and efficient conclusion:

- ➢ **Completing Project Deliverables as per contract**: -
    - ○ **Meticulous Deliverable Evaluation:** Carrying out a meticulous assessment of every project deliverable to guarantee they adhere to predetermined quality benchmarks and correspond with customer requirements.
    - ○ **Final Verification:** Engaging with pertinent stakeholders to validate the thoroughness and precision of project deliverables, guaranteeing the fulfillment of all contractual commitments.

- ➢ **Conducting Performance Evaluations: -**
  - o **Performance Assessment:** Carrying out a comprehensive assessment of the FGD systems adopted, by comparing the actual outcomes with pre-established success criteria and performance indicators.
  - o **Client input:** Requesting input from clients and end-users to assess their contentment with project results and pinpoint areas for improvement.

Lessons Learned and Continuous Improvement: -

- ✦ **Post-Implementation Review:**

Conducting a post-implementation review to highlight achievements, difficulties, and opportunities for improvement experienced during the project's duration.

- ➢ **Recording Lessons Acquired:** Recording significant lessons acquired, capturing valuable insights from both accomplishments and difficulties, and preserving this knowledge for future consultation.
- ➢ **Continuous Improvement strategy:** Formulating a strategy for ongoing enhancement by using insights gained from past experiences, therefore establishing and implementing optimal methods to improve future project outcomes.

Project closure is not only an end, but also a chance for reflection, enhancement, and knowledge dissemination. To promote project success and improve project management practices within the FGD domain, I will prioritize the completion of deliverables to high standards, conduct comprehensive performance evaluations, and establish a culture of continuous improvement through lessons learned.

The exploration of my journey several periods in the Indian corporate environment demonstrates a significant and progressive development, marked by certain stages that have impacted the strategies and choices of stakeholders. Now, let's explore each period in further depth:

Era of Hospitality

Indian power plants enthusiastically embraced foreign knowledge, especially in sophisticated technology such as Flue Gas Desulfurization (FGD), during this time. Foreign companies were welcomed and included in knowledge-sharing seminars to synchronize tender requirements with global standards. The main emphasis was on using international expertise to facilitate progress in project implementation and technology adoption.

Age of Handshake

Driven by a positive outlook and motivated by programs like "Make in India," EPC (Engineering, Procurement, and Construction) businesses invited Original Equipment Manufacturers (OEMs) to set up local manufacturing facilities. Forecasts of future company volumes resulted in obligations that sometimes strained profitability. Notwithstanding the early obstacles, efforts were made to establish partnerships with international technological companies in anticipation of future benefits and potential for expansion.

Era of Negotiation/Execution

The Negotiation/Execution phase served as a moment of truth for players as they struggled to establish their positions and navigate the complexities of the Indian market. The actual expenses associated with doing business in India became evident, often posing difficulties to the original estimates and anticipations. As parties tried to align their goals with the actual situation, negotiations were widespread. This resulted in strategic changes and alterations to the plan.

The Age of Get Out

Currently, stakeholders are experiencing financial limitations that originate from the initial planning stages. These limits are further intensified by changing government policies that prioritize indigenous ownership and control. The business environment has shifted from the "Make in India" approach to the "Make by Indian" approach, which highlights the importance of local production and self-reliance. Foreign partners face significant pressure to surrender control under unfavorable conditions, leading some to reassess their participation in the market and, in some instances, withdraw completely.

The transition from the Era of Welcome to the Era of Get Out represents the changing dynamics of globalization and regulatory changes in the Indian corporate environment. Every stage leaves a unique imprint, molding the tactics, choices, and future path of enterprises functioning in India. The ability of stakeholders to adjust to evolving conditions and overcome new obstacles will be vital in deciding their success in the dynamic Indian market.